Table of Contents

   I. Preface....................................2
   II. Dedications, by the students...............3
   III. Helminths...............................5
   IV. Insecta.................................12
   V. Mites and Ticks.........................20
   VI. Miscellaneous (Leeches and Lamprey)......27

I. Preface:

This book is the 4th in a series written by students in my Invertebrate Zoology and Parasitology classes, since 2012. In each case, I have made an offer to my students that they really couldn't refuse: "Either I stick you with a final exam, (and you *know* I can stick you with a final exam), OR you can write a children's book on <fill in the blank>". Not so surprisingly, the students in these upper-level classes choose the alternative of writing a book.

I have found that the experience of writing a book and drawing the illustrations for it is an exercise that engages the students far more effectively than writing (and then grading!) a final exam.

Some of the illustrations are accurate representatives of the organism being discussed, but some are quite imaginative (see *Enterobius vermicularis*, the pinworm, or *Dirofilaria immitis*, the heartworm.)

Other books in this series include the following:

Don't Get Sick, Stan!	ISBN 978-1469-944616
All Creatures Small and Smaller: The World of Invertebrates	ISBN 978-1475-295528
Oh, No, Not Moe Mosquito!	ISBN 978-1519-518910

I gratefully acknowledge Dr. Samantha Alperin, who has given the students in each class an introduction to the world of writing books for kids.

II. Dedications, by the students:

Student Name	Dedication
Stephanie Allen	Dave and Denese Allen, Robert Tworek. I dedicate this book to my parents, Dave and Denese Allen. I also dedicate this book to Robert Tworek because I know it will gross him out.
Maddie Belou	To my friends and family
Shakayla Bowdre	David Freeman and Sharon Bowdre
Samiha Elkhayyat	This is dedicated to my little brothers & sisters who I hope will enjoy reading this book as much as I enjoyed writing it.
Gil Erwin	Lance Silvis
John Trey Gillenwater	This book is dedicated to all the unknown parasites inhabiting the readers and authors; without them none of this would've been possible.
Nanday Kamanda	To mother, Stella Kamanda, and to Aunt and Uncle Gessie & Jibou Tucker
Peter Kinsella	To Lincoln Stoltz
Hsuan-Hsin (Abraham) Lin	To Paul & Hope Lin, Charity Lin
Sophiana Lindenberg	To Alen Vela, Lance Silvis, Zach Lindenberg, Steve Buscemi, Dr. Eisen
Andrew McKeever	To Lincoln Stoltz
Sara Owen	To Rocko the dog

Elizabeth Parr	To my parents, Rocio Parr & Bill Parr; to my sister, Virginia Parr; to my brother Will Parr; to my triathlon coach Heather Nichols & all the GREATEST professors at CBU, including Dr. Stan Eisen.
Courtney A. Pendergrass	To Dr. Eisen, the Parasitic Guru
Keane Reneé Prosser	I dedicate it to Dr. Eisen & the Christian Brothers University, who don't suck but made this all possible!
Maria Ruiz	I would like to dedicate this book to my siblings, so they can be aware of the parasites around the world and know what these things could do to them!
Lance Silvis	To Jesse Silvis, Steve Buscemi, Stan Eisen
Shane Talley	To Dr. Eisen, Dr. Henson, and Wilson Phillips
Sinead Tierney	To JT
Alen Vela	To Steve Buscemi, Stan Eisen, James Moore, Max Vela
Kelsie White	To all those who've been affected by parasitic diseases, Dr. Eisen, Stanley White Sr. And Tabatha White, Stanley White Jr., Debra Curry, Linnie Ellis and Sara White
Antoinette S. Wilson	I would like to dedicate this book to all of the people who have suffered from parasitic diseases and didn't make it. May you rest in peace. I would also like to dedicate it to Penny White.
Megan Wilson	Dedicated to my family, friends, and anyone who has suffered from a parasitic disease.
Lily Wong	To Dr. Eisen and to all the children who like blood-sucking parasites

III. Helminths

Pork Tapeworms: *Taenia solium*

Tapeworms are very long worms and can live in a person's intestines. They eat the food that goes into their digestive system. Tapeworms are also called cestodes and this one is called *Taenia solium*.

How do you get a Pork Tapeworm?

You get the pork tapeworm by eating the larvae in undercooked pork. The larvae enter your digestive system and grow in your intestines. They will then grow up and have more eggs while living inside of you. The eggs leave the body in the poop and they re-enter the environment to find more pigs.

However, you can also get a scary disease called **Cysticercosis**. The tapeworm can migrate throughout the body, like muscle and the brain.

Symptoms:

- Pain in your stomach
- Nausea or vomiting
- Headaches

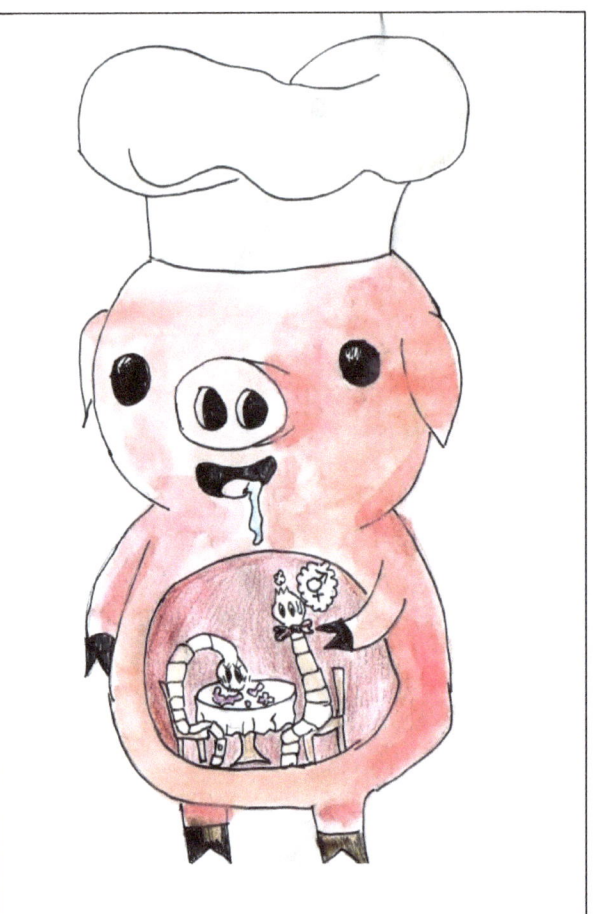

How do you prevent it?

<u>Always</u> have an adult help prepare meat, especially pork. It should be frozen for a period and then cooked to the proper temperature.

Chinese Liver Fluke: *Clonorchis sinensis*

The Chinese Liver Fluke is found in Korea, China, Taiwan, Vietnam, Japan, and Asian Russia.

How do you get infected with Chinese Liver Flukes?
    First of all, eggs are found in poop in the water. Snails then come along and eat the eggs. Next, the eggs grow in the snail and when they grow enough, they crawl out of the snail. They then go on to attack a fish and crawl into their skin. Finally, they enter animals or humans when they eat the fish infected with them. Once in the human, they travel into the bile ducts and hurt the host's liver. A Chinese liver fluke can produce 4000 eggs per day and live for 30 years! WOW!

The good news is that there are ways to prevent being infected with Chinese Liver Flukes. The best way is to not eat fish without cooking it or to freeze it at very cold temperatures. Cooking and freezing the fish works by killing the worms! If you do get infected with Chinese Liver Flukes, there is good news because you can be treated with anti-worm medications.

Encyclopedia of critters

Heartworms: *Dirofilaria immitis*

This is a worm that will live in the heart of dogs and cause it to become very sick by causing lung disease, heart failure, and other problems.

How does a dog get Heart Worms?

First, an animal infected with heartworms will have baby heartworms in their blood. Then, a mosquito will bite the sick animal and carry the worms and bite a pet. Once the mosquito bites a second pet, the heartworms grow up inside the new animal.

Sometimes, if the heartworms grow old enough, the adult worms will live in the right side of the pet's heart and make it harder for it to pump blood.

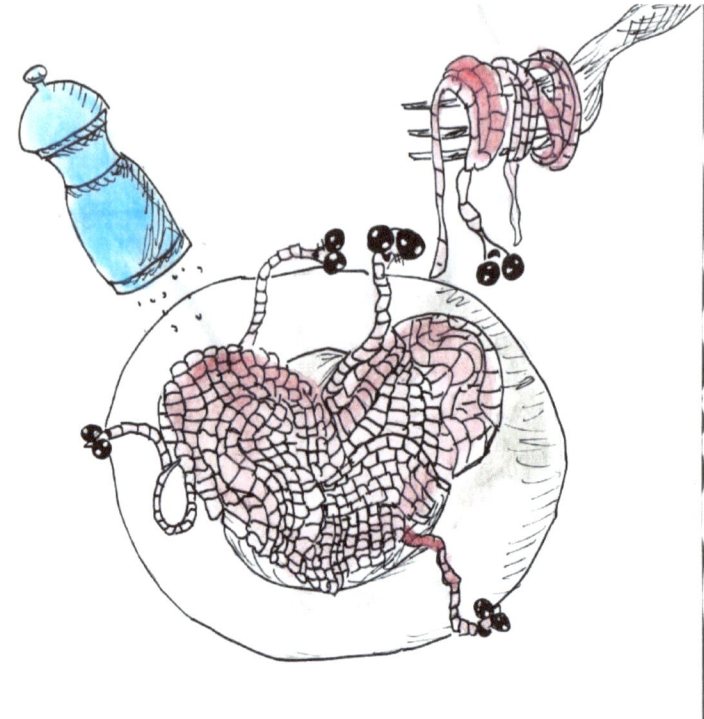

Pet owners can prevent heartworms by giving their pets medicine every month.

Pinworms: *Enterobius vermicularis*

It is a tiny white worm that get into the intestines and makes a person's bottom very itchy.

A person gets pinworms, when they accidentally eat the worm's eggs and they go into their intestines. The tiny white worms then hatch and the female worms will crawl to the exit and lay their new eggs. The eggs cause the person to itch a lot. The eggs in the bottom then get stuck under the nails and on the fingers. The eggs then go onto other surfaces that other people touch. This is how other people can get pinworms.

How do you prevent it?

If you have pinworms, avoid touching surfaces with dirty fingers, and don't put your hands in your mouth. Thoroughly clean your hands and fingernails, and make sure you clean them often.

Worm Therapy: How Worms Can Be Helpful

If worms can do all of these terrible things, then why would it ever be good to have worms?

Some individuals are infected with worms on purpose as a form of treatment; this is called Helminthic Therapy. Sometimes people's bodies try to attack themselves and worms can be used as a distraction. Worms can be used to treat diseases such as inflammatory bowel disease, asthma and multiple sclerosis.

How does it work?

Individuals are exposed to the larvae of hookworms. The patients are then checked closely by a doctor. The theory behind using worms as a treatment is that it will inhibit or decrease the body's immune response

Individuals have the risk of developing serious side effects over the course of the treatment. There is an increased risk for anemia, so iron supplements may be prescribed. Some other side effects are protein deficiencies, difficulty in thinking, and stunting of growth.

Although helminth infections have been shown to inhibit the immune system in mice and rats, studies with humans have not been conclusive. Nonetheless, the idea seems promising and there will likely be more research in the near future!

And here's an example of a therapeutic roundworm:

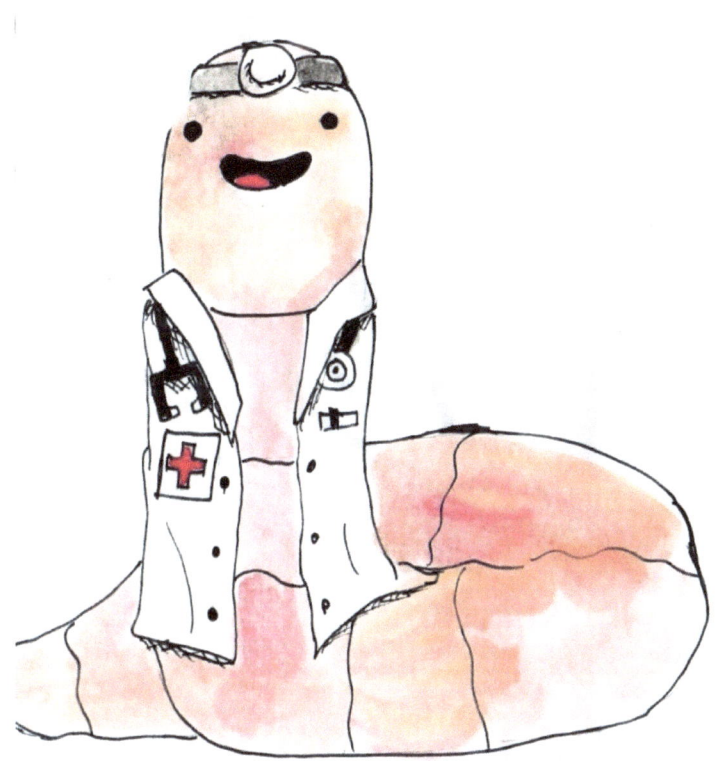

IV. Insecta

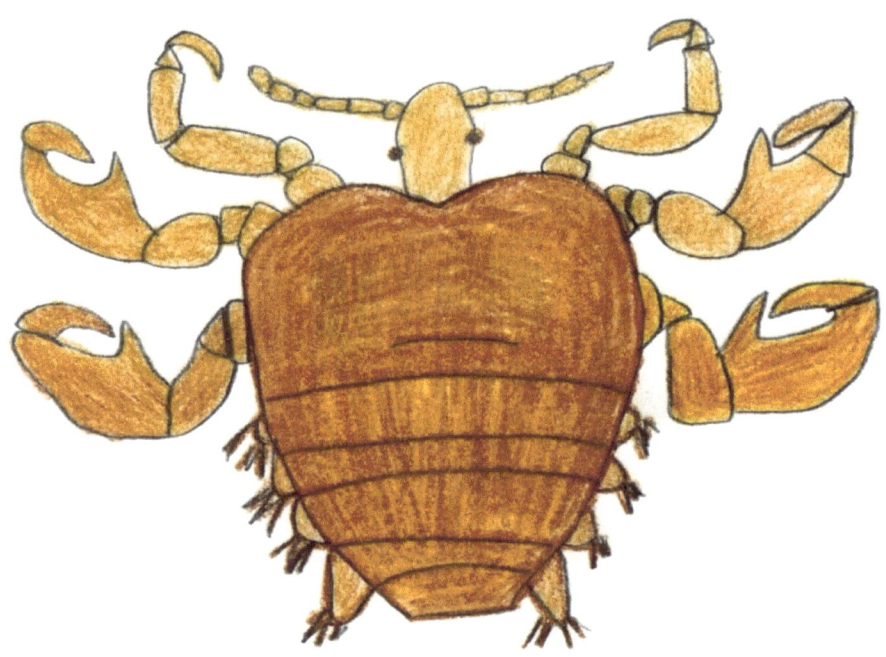

Black Flies: *Simulium* spp.

Black flies cause this disease called River Blindness because they carry a parasite that affects the eyes and skin.

Found: All around the world, but the disease is only a problem in tropical places. It usually takes a lot of bites for it to be very effective.

Weird enough, black flies aren't just black. They can be gray, brown, or yellow

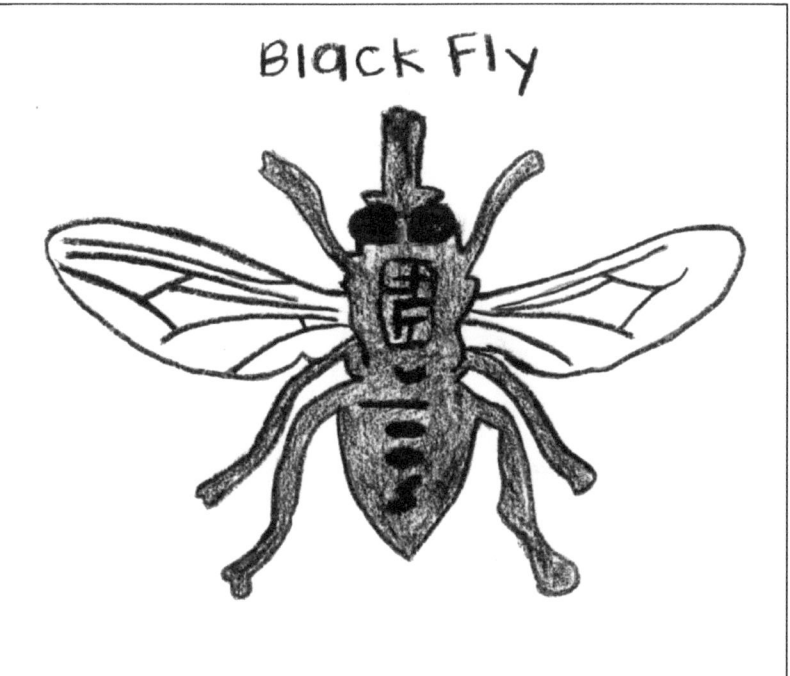

Flea: Pulex irritans

Fleas are wingless insects with a reddish-brown color and a long syphon called a proboscis made to pierce the skin and suck their host's blood. They are found anywhere there is people. Rats, dogs, and cats are common causes for a person to get fleas.

The species of flea *Xenopsylla cheopis* transmitted the Bubonic plague throughout Europe and wiped out 1/3 of Europe's population.

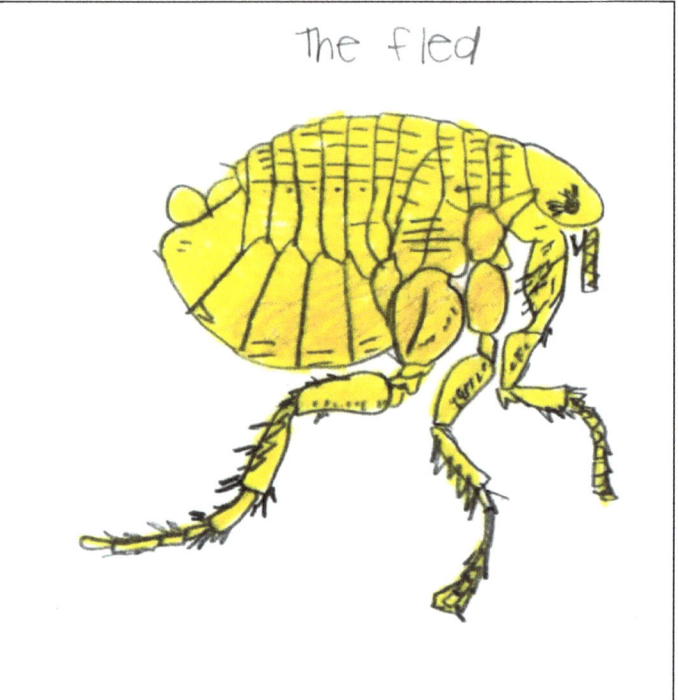

Tsetse fly: *Glossina* spp.

The tsetse fly is a large biting fly found in Africa, south of the Sahara desert.

The diseases it transmits are all caused by the genus *Trypanosoma,* and include Nagana (a form of trypanosomiasis, infects mainly cattle and horses) African sleeping sickness (infects humans, fatal if not treated).

Luckily there are ways to avoid getting bitten:

1. Wear thicker clothing that tsetse fly cannot bite through;

2. Neutral colored clothing, since the tsetse fly is attracted to bright colors;

3. Avoid bushes- where they rest during the hot temperatures of the day.

Assassin Bugs: *Rhodnius prolixus*

Rhodnius prolixus is a blood-sucking, reddish-brown bug that can be over an inch long. This blood sucker bites when a person is sleeping. They can be found everywhere but are mostly located in South America.

Assassin bugs transmit Chagas' disease which is due to the protozoan parasite called T. *cruzi*. The disease can cause brief, mild illnesses or chronic severe illnesses. If left untreated for too long, severe heart and digestive problems can occur. Sucks right?!

Mosquitos: Family Culicidae

Mosquitoes have one set of wings, and a proboscis for feeding. It has antennae that detect hosts and breeding sites where females can lay their eggs, male antennae are bushier than female antennae. They are found everywhere in the world except Antarctica, Iceland, New Caledonia, the Central Pacific Islands, and Seychelles.

They transmit numerous viral and parasitic diseases, including dengue fever, West Nile Virus, malaria, yellow fever, viral encephalitis, and Chikungunya virus.

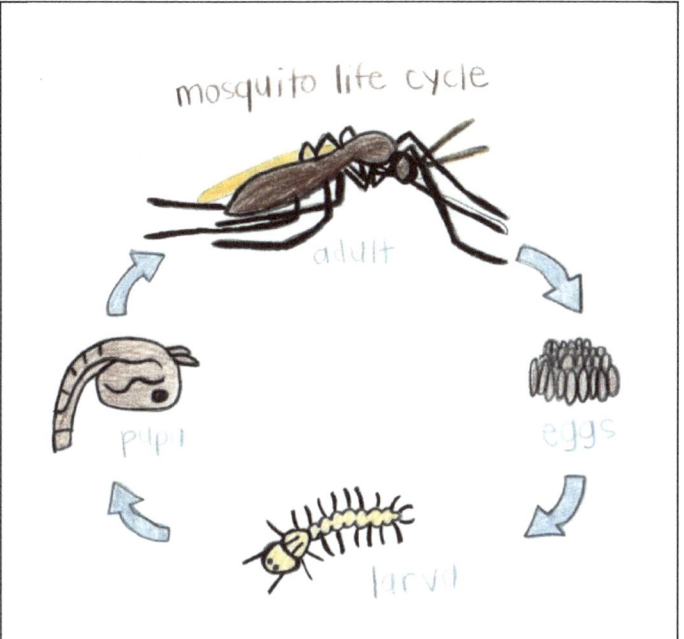

Body Lice: *Pediculus humanus humanus*

Also known as *Pediculus humanus humanus*, lice live on human skin biting you and feasting on your blood!

Body lice can transmit a form of typhus fever. When a louse bites you, itchy red bumps or rashes will appear on your skin.

A special shampoo or high heat can be used to get rid of these creepy crawlies!

Bed Bugs: *Cimex lectularis*

Bed bugs are typically a reddish-brown color with a flat, oval body. They also only feed on blood. They can be found all over the world, and can hide inside walls.

Thankfully they don't carry diseases, but they bite—A LOT and can be really hard to get rid of. The thing that works best to get rid of them? Extreme heat!

The Bed Bug

V. Mites and Ticks

The Rocky Mountain Wood Tick: *Dermacentor andersoni*

The Rocky Mountain Wood Tick is a small insect that attaches onto the skin of some vertebrates and feeds on their blood.

It lives in the northwestern United States along the Rocky Mountains and some parts of Canada. Its common habitat is in shrublands, wooded areas, and open grasslands. They only get from about 5 to 6 mm long.

A viral infection known as Colorado tick fever is transmitted by the Rocky Mountain Wood Tick. You can get Colorado tick fever in the hot summer, along the Rocky Mountains.

Don't want your blood sucked?

Prevent ticks from getting on you by wearing long sleeves and pants, using bug spray, avoiding tall grass, and performing tick checks and removing them immediately, if seen. BUT if you've seen a tick on you and you've experienced a red rash, fever, chills, headaches, muscle aches, loss of appetite, body aches or weakness, go see a doctor.

The Black-legged Tick: *Ixodes scapularis*

If you have ever taking a hike up a mountain or gone camping you might have taken an uninvited guest home with you. The guest has black legs and is called *Ixodes scapularis* or the black-legged ticks. They leave a red circle near the area they bit you as a gift. They are usually found in wooded areas and North Eastern United States.

The females have eight black legs, long flat body and are dark red-brown color. They are like vampires because they suck the blood of all types of animals and get bigger.

While they suck blood, the ticks inject a small germ called *Borrelia burgdorferi*, that causes Lyme disease. Lyme disease is like a ninja because it hides in your body until the person is at their weakest, then it attacks. When this happens, a person can lose their memory, have joint pain and heart problems. If you ever see a red spot or feel ill, tell a parent or an older adult so they can take you to the doctor.

The Blue Tick: *Boophilus decoloratus*

If you've ever gone to a farm or been to a zoo, chances are you've seen cow. If you have ever gotten really close to them and noticed many tiny black spots, those could be ticks! If they are, then those ticks are most likely called *Boophilus*.

These ticks are found in the tropical and subtropical parts of Africa, Asia, and America. These ticks are black and very small. Female ticks are larger. Whenever the female ticks need a place for their babies to sleep, they attach to cattle.

However, when these ticks live on cattle they can spread diseases such as Babesiosis. This disease can make the cow feel weak, lose weight, and even pass away. That's why you must take care of cows or any animals you own.

If the cow has the disease it is treated until it feels better and it is given the same kind of shot you would get at the doctor, a vaccination. To prevent your animal from getting this disease in the future and for it to be healthy, make sure you take it to the vet for regular check-ups and shots.

Encyclopedia of critters

The Brown Dog Tick: *Rhipicephalus sanguineus*

For many centuries, dogs have been man's best friend. Well, if you have a pet dog in your house then you should be on the lookout for ticks, like the brown dog tick, that may harm your little friends.

Dogs can be found almost anywhere, which means these brown dog ticks are too! But they are more abundant in the Southern part of the United States. The fully grown adults are found on dogs, but the larva stages are found on different types of rodents.

Some cool facts about these little creatures:

They are red-brownish color and they can lay about 5000 eggs when they reproduce. They also suck the blood of your doggy friends.

They can also transmit diseases to dogs that may cause lameness or fever and other symptoms that may not be noticeable. Make sure your doggy friend always get regular check-up from his/her veterinarian.

The Lone Star Tick: *Ambylomma americanum*

Have you ever felt a small bump on your puppy when you were petting her? If so, this might be a tick. One tick has a spot on his back, and it is called the lone star tick.

This species or type of tick is found in Maine and throughout the southeastern and eastern states. Ticks have 8 legs like spiders do and can be many different colors. What's special about this tick is that the females have a white spot on their back. When a tick gets hungry it will bite an animal or human and suck its blood. Their favorites are turkey or deer blood.

Ticks can transmit many different diseases, including Rocky Mountain Spotted Fever. This disease is caused by a bacterium the tick injects when biting a human or animal. When a person has this disease, they might have a headache, a rash and pain. It is important to tell a grownup if you notice after finding a tick on your body.

If the doctor thinks the tick gave you Rocky Mountain Spotted Fever, he/she will give you some medicine called an antibiotic.

Dust Mites: *Dermatophagoides farina*

These mites live in dust and can cause allergic reactions such as wheezing and asthma in humans. You can try to get rid of them in your house by changing your bed sheets regularly! If these mites happen to live in your home, don't worry. They are not parasitic and they can't burrow under your skin!

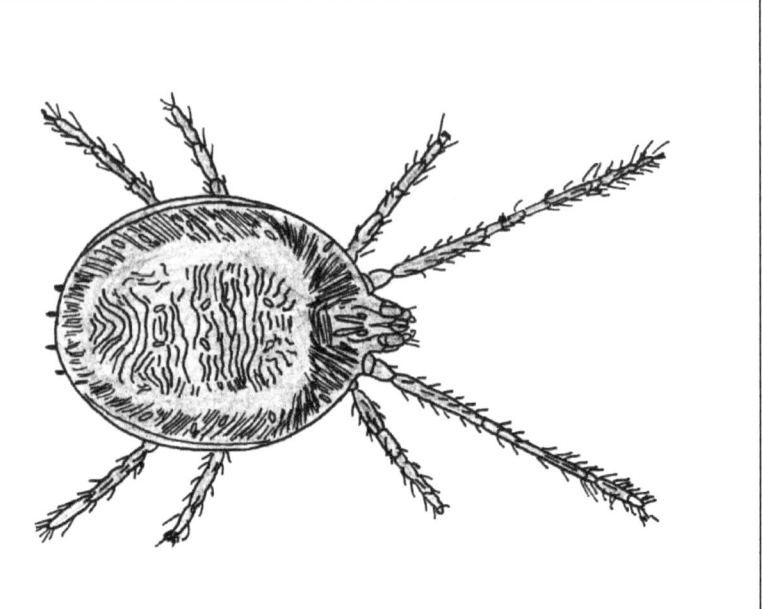

IV. Miscellaneous (Leeches and Lamprey)

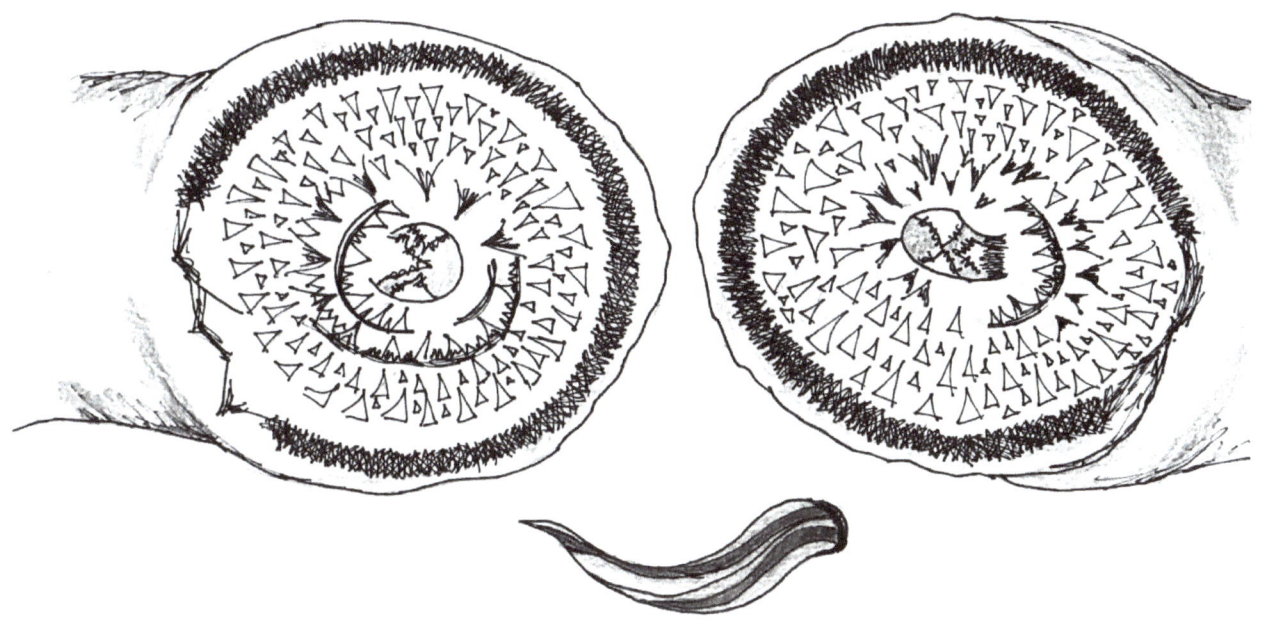

Medicinal Leeches: *Hirudo medicinalis*

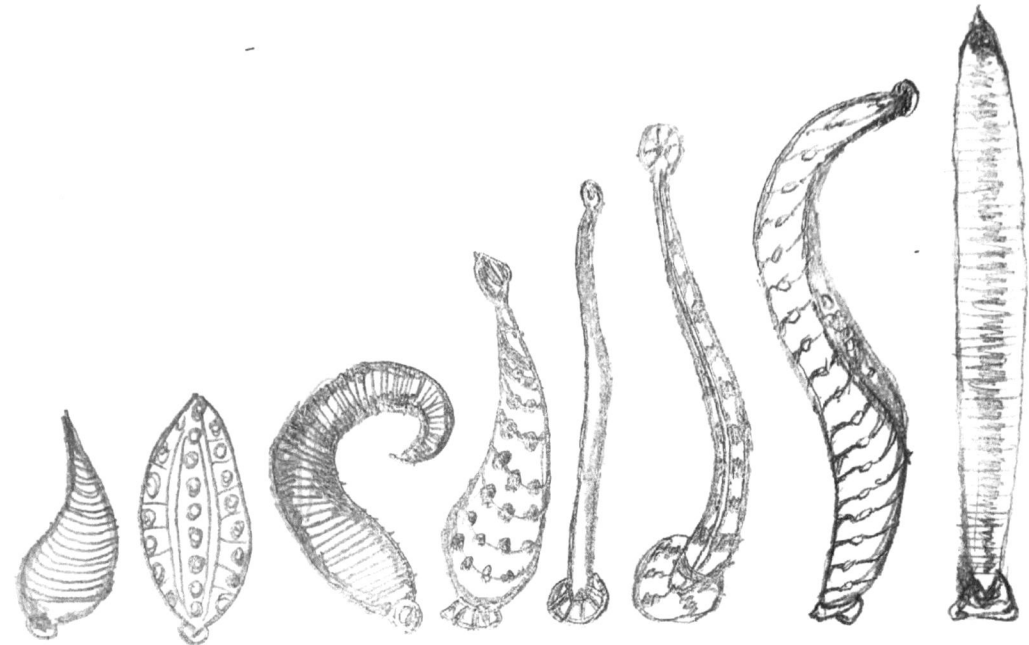

Leeches are characterized as these flat, wide, segmented creatures that have suction cups at the end of their bodies. They come in different colors, such as black, brown, or green. Their skin can have spot and stripe patterns on them, or just be plain. These creatures come in all sorts of sizes. Some have been found to be an inch while others can be 10 inches. The species were known to feed on fish, animals, and humans have sharp teeth. Leeches from different species can be found all over the world.

Medieval and Renaissance Uses

For thousands of years, leeches have been used as a treatment for a variety of illnesses like plague, small pox, and seizures. The first recorded history of medicinal leech use was in ancient Egypt, spreading to Rome, Asia, and finally into Europe.

During the Middle Ages and Renaissance period, medicinal leeches were put on the body near the area that was suspected to be causing the illness. So if a person was coughing very badly, doctors would put leeches on the chest near the lungs. People believed that the leech would then suck out the "bad blood" from that area, and cure the sick person. Leeches can suck anywhere from 5-10 ml of blood! That can be five times the leech's weight.

Hirudotherapy

Leeches are used to treat venous insufficiency in surgical free flaps and most commonly in finger and ear replantation. Saliva from leeches contains agents which medically serve as anticoagulants. These chemicals include hirudin, an anticoagulant peptide, and calin, a platelet adhesion inhibitor, which both work to prevent or dissolve blood clots and thrombi.

In this manner, leeches provide the means to remove stagnant blood, allowing for the dilation of blood vessels and an increase blood flow in the area being treated, until replantation can occur. The number of treatments for this therapy varies depending on the degree of congestion and sessions can last between 30-90 minutes.

The leeches used in this procedure are raised in strict pharmacy conditions as to prevent bacterial infections and patients are closely monitored with complete blood cell counts to prevent any excess blood loss.

Sea Lampreys: *Petromyzon marinus*

The sea lamprey is known to be one of the most primitive vertebrate species. They are very successful in the water when it comes to latching onto their prey due to the anatomy of their mouth. Characteristics such as sharp, horny teeth and suction-cup, disc shaped mouth make life easier for this species. They've caused a great decrease in populations among important fishes, such as whitefish, trout, and perch, throughout well-known lakes. Organizations are equipped to control the sea lampreys in such areas.

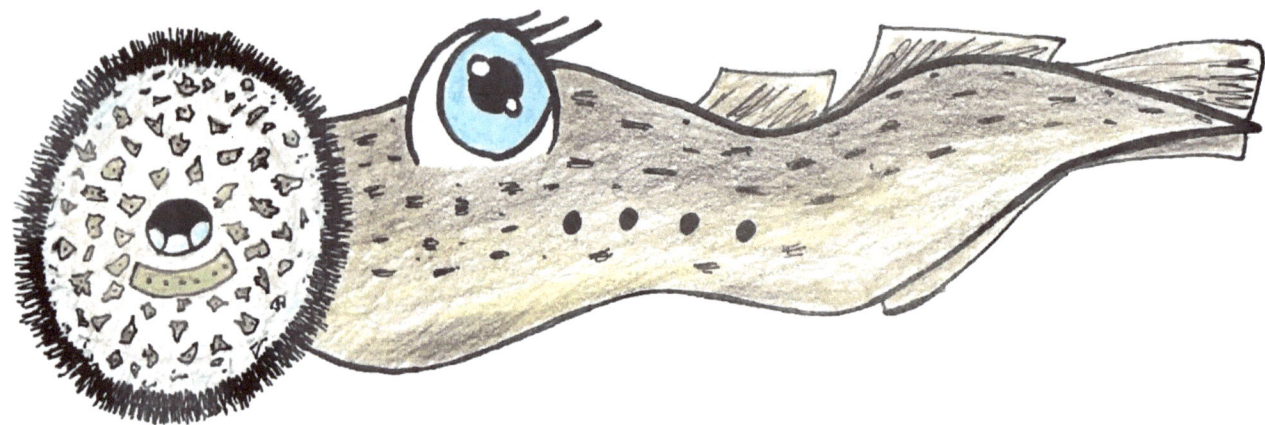

Migration into the Great Lakes

The sea lamprey is one of the most destructive invasive species in the Great Lakes region. These parasitic fish are present in all five Great Lakes and have caused a significant decline in native open water fish populations in the upper three Great Lakes.

Encyclopedia of critters

Like many native fish, sea lampreys spend their adult lives in the open lakes but use tributaries for spawning. They spawn in gravel areas in the flowing water of the tributaries of the Great Lakes. After hatching, lamprey larvae prefer the soft sediment of these rivers and filter feed on algae, detritus, and microscopic organisms and materials. As they mature into adults, the larvae migrate to the cooler waters of the five Great Lakes. Over their 12-20-month adult feeding period, each adult sea lamprey can kill up to 18 kilograms (40 pounds) of fish

They have an important place in their native ecosystems. Sea lamprey attach to lake trout, brown trout, lake sturgeon, lake-whitefish, ciscoes, burbot, walleye, catfish, and Pacific salmonids in the Great Lakes. Many of these fish are killed during the attack, while others die from indirect effects like infection of wounds left by lamprey. The loss of fish to sea lamprey is both ecologically and economically devastating for the Great Lakes region. Ecologically, large fish play an important role as top predators, and their absence can lead to a significant increase in prey species, which starts a domino effect on other parts of the food chain. The decline of predators resulting from sea lamprey can also facilitate the invasion of prey species.

Effects on Commercial and Sport Fishing

The introduction of the sea lamprey has caused a great decline in fish populations worldwide. Walleye and lake trout are two victims of these blood suckers. They have also been the reason for extinctions of certain cisco species. Additionally, these species have effected commercial fisheries, especially in the 1950s and 1960s in the Great Lakes.

It has ruined commercial lake trout fisheries, such as Huron, making millions of pounds to a fishery not worth exploring by fishermen. Not only do they affect fish, but the economy has also been a victim of the sea lamprey.

It's been reported that the United States more than $12 million each year on trout restoration programs to treat the damage caused by sea lampreys.

The reason they are successful and doing so much damage is because of their way of getting rid of their prey. Not only can they eliminate the prey by direct loss of blood and tissue but they also expose them of infections due to the open wounds they cause. These open wounds can even attract other predators. So it makes sense how they are able to affect fish populations and the environment so heavily when we understand their course of action for a meal.

www.ingramcontent.com/pod-product-compliance
Lightning Source LLC
Chambersburg PA
CBHW051819210526
45473CB00005B/1657